SAE EDGE™
RESEARCH REPORT

Unsettled Topics on Nondestructive Testing of Additively Manufactured Parts in the Mobility Industry

Kevin T. Slattery, D.Sc
The Barnes Global Advisors

EDGE DEVELOPMENT TEAM

Anton du Plessis, Ph.D, *Stellenbosch University*
Michael Goerlik, Ph.D, *Federal Aviation Administration*
Dan Johns, *BEAMIT SPA Additive Manufacturing Technologies*
Mick Maher, *Maher & Associates LLC*
Ankit Saharan, Ph.D, *EOS*
Jim Williams, Ph.D, *The Ohio State University (Ret.)*
Dylan Yazbeck, *Jesse Garant Metrology Center*

SAE INTERNATIONAL®
Warrendale, Pennsylvania, USA

About the Publisher

SAE International® is a global association of more than 130,000 engineers and related technical experts in the aerospace, automotive and commercial-vehicle industries. Our core competencies are life-long learning and voluntary consensus standards development. Visit sae.org

SAE EDGE™ Research Report Disclaimer

SAE EDGE™ Research Reports focus on topics that are dynamic, in which knowledge is incomplete, and which have yet to be standardized. They represent the collective wisdom of a group of experts and serve as a practical guide to the reader in understanding unsettled subject matter. They are not meant to provide a recommended practice or protocol. The experts have assembled as a community of practitioners to contribute and collectivize their thoughts and points of view; these are not the positions of the institutions or businesses with which they are affiliated, nor is one contributor's perspective advanced over other contributors. SAE EDGE™ Research Reports are the property of SAE International and SAE alone is responsible for their content.

About This Publication

SAE EDGE™ Research Reports provide state-of-the-art and state-of-industry examinations of the most significant topics in mobility engineering. SAE EDGE™ contributors are experts from research, academia, and industry who have come together to explore and define the most critical advancements, challenges, and future direction in areas such as vehicle automation, unmanned aircraft, cybersecurity, advanced propulsion, advanced manufacturing, Internet of Things, and connectivity.

Related Resources

SAE MOBILUS® Advanced Manufacturing Knowledge Hub
https://saemobilus.sae.org/advanced-manufacturing/

SAE Team

Frank Menchaca, Chief Growth Officer
Michael Thompson, Director of Standards, Information and Research Publications
Monica Nogueira, Director of Content Acquisition
Beth Ellen Dibeler, Product Manager
William Kucinski, Managing Technical Editor

Printed in USA

EPR2020017
ISSN 2640-3536
e-ISSN 2640-3544
ISBN 978-1-4686-0240-1

To purchase bulk quantities, please contact: SAE Customer Service

E-mail: CustomerService@sae.org
Phone: 877-606-7323 (*inside USA and Canada*)
+1-724-776-4970 (*outside USA*)
Fax: +1-724-776-0790

https://www.sae.org/publications/edge-research-reports

About the Editor

Andy Kunkel—Triad Production Group

Kevin T. Slattery, D.Sc—Kevin is a Principal ADDvisor® at The Barnes Global Advisors (formerly The Barnes Group Advisors). His primary expertise is in Metallic Additive and Metals Manufacturing, focusing on test program development, process and product verification, qualification, and certification. He has supported over 25 clients on five continents throughout the entire additive manufacturing value chain—from raw material to finished components.

Kevin was previously the Chief Scientist for Additive Manufacturing at Boeing Research and Technology (BR&T). He was responsible for developing and integrating the technology roadmaps and development plans for metallics additive manufacturing for the entire company, along with building and leading a multi-skilled team to execute and deliver the technology throughout the enterprise. Prior to that, he was the Chief Scientist for Metals, Ceramics, and Mechanical Systems at BR&T, with the responsibility for portfolio development and coordination, while executing the additive manufacturing portion.

He served as the Division Chief Engineer for the US Navy and US Air Force fighter aircraft and US Army rotorcraft in Boeing's Military Sustainment organization. From 1997 to 2012, he was on the BR&T Metals Team as a researcher and senior manager, where he primarily developed advanced low-cost titanium processing technologies supporting all Boeing products. He was the technical and programmatic lead in implementing the first aerospace metal-additive manufactured structural aircraft components for both spares and production, with five other first-in-the industry technology implementations.

He began his career at McDonnell Douglas (now Boeing) as a nondestructive testing engineer, where he developed inspection technologies for metallic and composite components, along with integrating impact of discontinuities with the acceptance criteria for carbon/epoxy composites.

Dr. Slattery holds BS and MS degrees in Metallurgical Engineering from University of Missouri-Rolla (now Missouri S&T) and a D.Sc in Material Science and Engineering from Washington University in St. Louis. He currently holds 36 US patents, with another 15 applications pending; along with 36 significant publications and conference presentations.

contents

Atosan/Shutterstock.com

Unsettled Topics on Nondestructive Testing of Additively Manufactured Parts in the Mobility Industry

Abstract

Additive manufacturing (AM) technology, also known as 3D printing, has transitioned from concepts and prototypes to part-for-part substitution and the creation of unique AM-specific part geometries. These AM applications are increasingly present in demanding fields such as medicine and aerospace. Generally, however, the applications are still primarily driven by thermal, stiffness, corrosion, and static loading conditions. In order to move to the next levels of structural significance (durability and damage tolerance), the AM technologies and components will need to reliably demonstrate freedom from inherent discontinuities that degrade durability to the point of preventing consideration in fatigue environments. This also includes freedom from rogue discontinuities that have the same impact on damage tolerance. Not only will the processes themselves need to be stable, robust, and repeatable enough to minimize the probability of such discontinuities occurring in the first place, but nondestructive testing (NDT) technology and methods will need to be sufficiently capable and reliable to ensure (within a reasonable level of risk) that such discontinuities will be detected to prevent the components from being accepted for use. Finally, while not the focus of this report, they will also be needed to inspect in-service components for damage and the start or growth of fatigue cracks.

As the second installment of a six-part series of SAE EDGE™ Research Reports on AM, this report discusses the need, challenges, technologies, and opportunities for NDT in AM. While it will focus on metallic components and technologies, it will also introduce ceramic and polymer composite components, as these technologies are being developed and implemented in increasingly mission-critical, if not durability and damage tolerance-critical applications.

NOTE: SAE EDGE™ Research Reports are intended to identify and illuminate key issues in emerging, but still unsettled, technologies of interest to the mobility industry. The goal of SAE EDGE™ Research Reports is to stimulate discussion and work in the hope of promoting and speeding resolution of identified issues. SAE EDGE™ Research Reports are not intended to resolve the challenges they identify or close any topic to further scrutiny.

KEVIN T. SLATTERY, D.Sc
The Barnes Global Advisors

Edge Development Team

Anton du Plessis, Ph.D, *Stellenbosch University*
Michael Goerlik, Ph.D, *Federal Aviation Administration*
Dan Johns, *BEAMIT SPA Additive Manufacturing Technologies*
Mick Maher, *Maher & Associates LLC*
Ankit Saharan, Ph.D, *EOS*
Jim Williams, Ph.D, *The Ohio State University (Ret.)*
Dylan Yazbeck, *Jesse Garant Metrology Center*

ISSN 2640-3536

A Note on Terminology

- **Post-Build NDT**—Inspections performed after a part or material product is made. This will be referred to in the report as NDT, and includes typical methods such as radiographic, ultrasonic, eddy current, penetrant, magnetic particle, resonant, and thermographic methods.
- **Process Monitoring**—Monitoring the inputs and responses of a process while it is being performed to ensure that it is operating within limits.
- **In Situ NDT**—Inspections performed while a part is being made using AM. This may use the same sensors and analysis methods similar to process monitoring, with the difference being that the output of the analysis is a region or a part that contains or does not contain a physical discontinuity and its approximate size.
- **Process Control**—Using data from process monitoring to change the inputs to a process to keep it within desired limits.
- **Physical Discontinuity**—A unintentional condition in a specific location of a material or part where the material is discontinuous, and load cannot be transferred. Typical physical discontinuities are cracks, pores, lack of fusion, and inclusions.
- **Defect**—A physical discontinuity that compromises the ability of a part to meet design intent.

Introduction

State of the Industry

Additive manufacturing (AM) is transitioning from technologies to make prototypes with only geometric requirements to producing performance-critical parts by the thousands for a variety of industries (also called "serial production"). Assuring the reliability of the parts produced will be critical to a successful transition. While process control and destructive mechanical and micrographic testing of witness coupons will play a role, key to this transition will be the ability to perform nondestructive testing (NDT) of both the surface and the interior (volumetric) regions of the part. NDT has been extensively used in the aerospace, shipbuilding, chemical, oil and gas, and energy fields (with a lesser role in the rest of the mobility industry) over the last 80 years. During this time, an impressive number of new materials and manufacturing processes have been successfully introduced for ever-demanding applications. The success can be noted in how rarely a part fails due to deficiency in material and processing. As AM technologies are introduced into more challenging applications, a similar level of product assurance will be needed. The different way in which parts are built means that the selection, development, and validation of NDT methods for AM processes represents an unsettled topic. Additionally, because AM does not have wrought metallurgy's long history of reliability, a second unsettled topic is determining the need for NDT in mobility industries, like automotive, that do not make extensive use of NDT.

NDT primarily assesses parts by looking for physical discontinuities, both internal and external (e.g., pores, cracks, laps, etc.). In the case of volumetric inspections, energy is transmitted through the part; differences in how that energy is transmitted through the part is indicative of an internal discontinuity. The difficulty of energy introduction and interpretation is directly proportional to the geometric complexity of the part. Since complex part geometry is one of the enabled benefits of AM, this task often results in difficulty for NDT practitioners. In the case of surface inspections, a relatively smooth surface is required, which many AM processes, due to their layer-by-layer build method or the use of fused material, do not produce. Additionally, the complexity afforded by AM in terms of internal holes and passageways makes viewing the surfaces even more challenging. This leads to the third unsettled area, the ability of conventional NDT technologies to inspect AM parts to the desired level of sensitivity.

While the rougher surfaces and complex geometries of AM parts make NDT more challenging, AM has some unique attributes that potentially help with product assurance. The first of these is that parts designed for AM tend to be thinner than conventional parts, so that accessing the internal features is less difficult. The second is that, by definition, AM parts are built in relatively thin layers. This means that discontinuities from the fusing process (e.g., pores, lack of fusion, etc.) could potentially be detected as the part is built with little geometric complexity or intervening material to complicate inspection. Unfortunately, the methods for performing these inspections, referred to as in situ NDT, are relatively new technologies, such that few have been validated to the degree seen in conventional NDT. This represents the fourth unsettled area, which is the ability of new, AM-specific NDT technologies to perform inspections to the desired level of sensitivity and confidence.

In summary, the four unsettled areas of NDT for AM parts in the mobility industry are

1. The selection, development, and validation of NDT methods for AM processes
2. Determining the need for NDT in mobility industries that do not make extensive use of NDT
3. The ability of conventional NDT technologies to inspect AM parts to the desired level of sensitivity
4. The ability of new AM-specific NDT to perform inspections to the desired level of sensitivity and confidence of detection

In this SAE EDGE Research Report, we will cover the AM processes and materials being considered for more demanding applications in the mobility industries, describe the challenges and opportunities that are unsettled in terms of the application of NDT, and finally where NDT of AM is headed and what its capability needs to be in order for the technology to make its full impact on the mobility industry.

Additive Manufacturing Processes and Materials for More Demanding Applications and Discontinuity Types

The Need to Detect Physical Discontinuities

Industry in general, and the mobility industry in particular, need to produce parts without physical discontinuities. This is because physical discontinuities above a certain size are defects that lead to premature failure of a part, leading to the loss of the system or requiring unplanned maintenance of a system. The mobility industry has a greater need to be free of discontinuities due to the following

- The need to minimize weight for efficiency means lower design margins.
- The cyclic (fatigue) loading mobile systems experience mean that small discontinuities in as-produced components that can grow to become cracks that lead to early replacement or failure.
- The consequence of failure due to remoteness from maintenance or being airborne or seaborne is higher than for nonmobile systems.

The best way to create parts without discontinuities is to have robust manufacturing processes that operate well within their limits, which is the approach used by most of industry, even the mobility industry. The majority of metal parts in the world are made from either flat-rolled (e.g., sheet, coil, band, thin plate), long-rolled (e.g., bar, rails, some pipe, structural shapes), or wrought products, either extruded or forged. These products are almost entirely free from discontinuities because of the cleanliness of the input stock, but more importantly, the large amount of reduction during hot working will break up inclusions or heal pores or cracks, making for an inherently robust process. Monitoring and control methods are used to ensure that a process stays well within its limits, especially ones that are less robust, such as automated welding.

NDT is used when a combination of the probability of a discontinuity occurring, the probability of said discontinuity compromising performance, and the consequence of compromised performance exceed a threshold of acceptability. Examples include

- Manually welded pipes in the chemical industry
- Large sand castings in an industry where the consequence of failure is significant (but not in small die castings in the same industry, where the probability of a discontinuity is higher, but the consequence is lower)
- Rolled aluminum plate in the aerospace industry where discontinuities almost never occur, but where the consequence would require unplanned maintenance and downtime for an aircraft

The first two examples are processes where a critical process parameter (e.g., manually following a groove, pouring the metal down the sprue) is quite instantaneous and the ability to perform process monitoring and control are limited. In the third example, some portion of the process (e.g., gap between rolls, plate temperature, etc.) can be carefully monitored and controlled, while others—such as the presence of an inclusion in the initial ingot—cannot. These examples also illustrate the different roles of NDT. In the first two, the only purpose of performing NDT is to provide assurance that a defect does not exist in the weldment or the sand casting. In the third example, not only is the goal to provide assurance that a defect does not exist, but it also as an additional demonstration that a process is under control.

As a new manufacturing technology, AM has a benefit over older manufacturing processes in that, as an automated layering process that runs for minutes and hours, it better lends itself to process monitoring and control. The other benefit of AM is that—in processes where parts are built in thin and usually have visible and accessible layers—there are more options to monitor the process and verify a lack of discontinuities (in situ NDT) than there are for 2,000 kg blocks of metal that are forged at 1000°C. A final benefit is that building parts in thin layers tends to limit the upper size of a discontinuity.

The drawback to AM, however, is that as a new manufacturing technology, there is no decades-long

- History of recognizing which process variables are key to robustness
- Demonstration of the ability to control the processes in real time and the appropriate variation limits
- Demonstrated understanding of which of the hundreds of process variables are key to the level of confidence that some industries require

Additionally, knowledge of problems in the past with conventional technologies has resulted in greater scrutiny being applied to AM at the same level of development. Note that in spite of the fact that the world's airplane fleets have billions of flight hours with little if any failures in titanium airframe forgings, all forgings still receive 100% volumetric NDT (ultrasonic) and 100% surface NDT (penetrant). Thus, even when AM reaches the point where there is a high level of confidence that a process is capable of creating parts that are free of defects, some form of NDT will be necessary, especially in applications where either safety or costly downtime is involved.

This brings us to the first unsettled issue regarding AM NDT, which is what kind of physical discontinuities can an AM process produce, and how does one determine if and when they need to perform NDT to assure freedom from discontinuities that are defects? In the following paragraphs, the relationship between materials, processes, and potential discontinuities and defects will be discussed, along with a recommended approach to determining if NDT is needed. We will use the seven (7) different process categories as defined by the ASTM F42 committee, noting that the following is not intended to be an exhaustive list of all material/process/discontinuity combinations and their causes, but rather sufficient to aid a practitioner in considering the possibilities and risks.

Ceramics

Binder Jet Printing, Material Jetting, and Vat Photopolymerization In terms of discontinuities, binder jet printing (BJP, Figure 1), material jetting (MJ, Figure 2), and vat photopolymerization (VP, Figure 3) are quite similar. Additionally, they are similar to ceramics produced using conventional processing. That is because all three processes print a green shape that is a mixture of the ceramic and a binder. This binder is then driven off either before sintering, or as part of the sintering operation that densifies and strengthens the ceramic part. Unless full density is achieved by a final hot isostatic pressing (HIP) or infiltration process, the final part will have porosity in it. Therefore, the primary discontinuities of concern are

- Cracks, which can be due to thermal shock during cool down, or a portion of the build slumping and cracking during sintering
- Pores larger than a certain size, which could impact mechanical or thermal properties—these can be due to binder rapidly evaporating and driving the ceramic out of a region or from an initial inhomogeneity in the ceramic powder
- Undesirable phases from either failure to remove binder that would react during sintering or incomplete reactions during sintering; depending on their size and interaction with the desirable phases, these may or may not be a physical discontinuity

FIGURE 1. Binder jet printing (BJP).

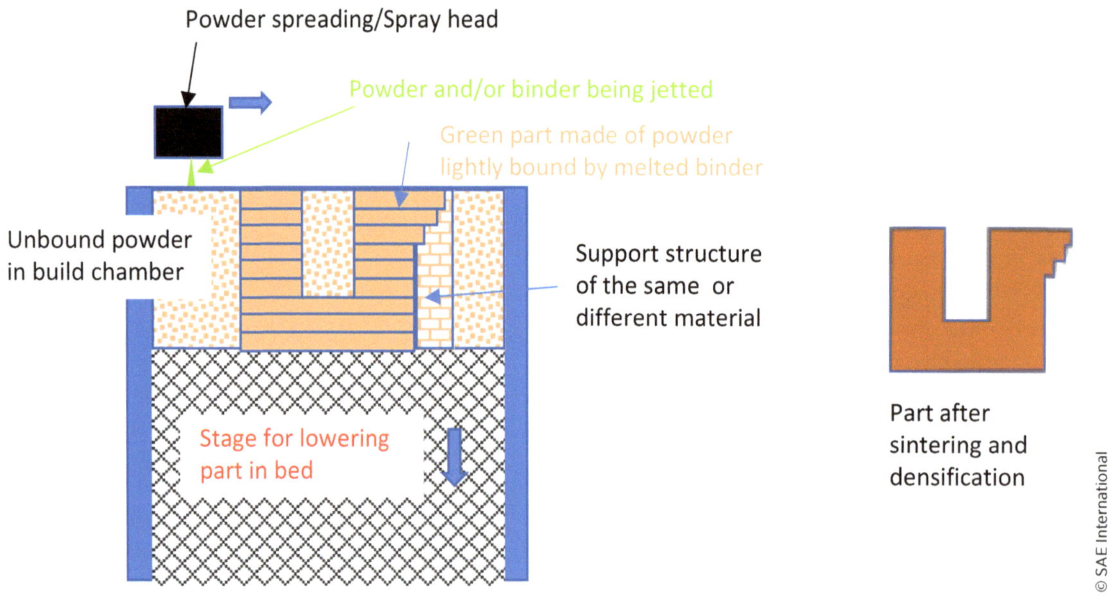

FIGURE 2. Material jetting (MJ).

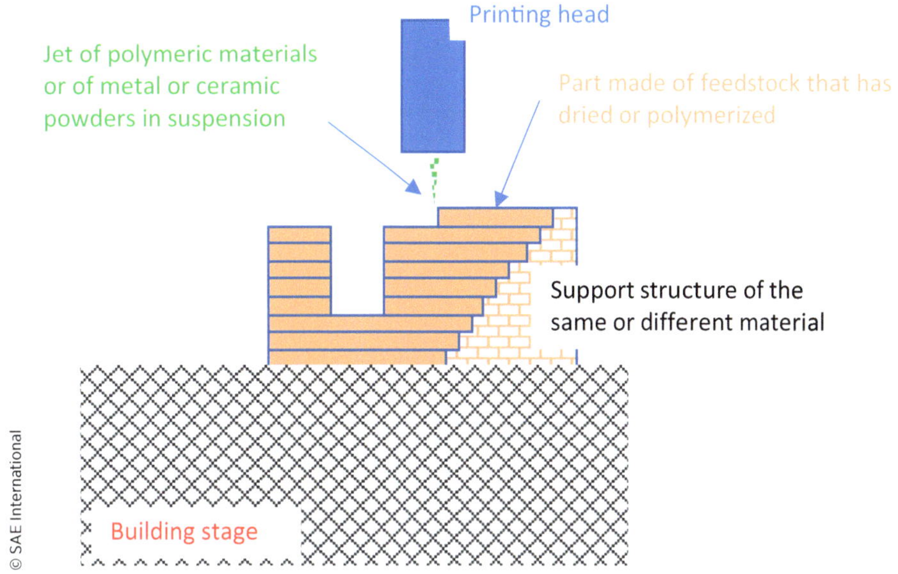

FIGURE 3. Vat photopolymerization (VP).

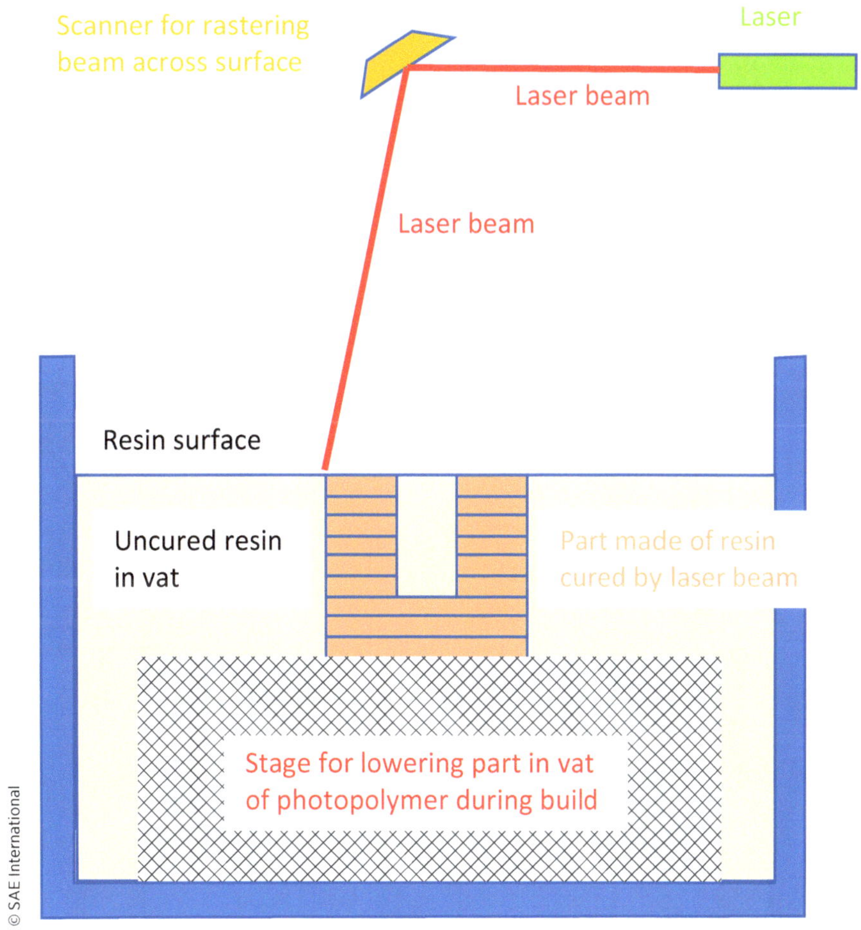

Finally, because of the nature of manufacturing, the orientation of these discontinuities can be considered random.

Polymeric Composites

Material Extrusion For the purposes of this SAE EDGE Research Report, we will focus on material extrusion (ME) of polymeric composites that have oriented fibers in them, while also mentioning other polymeric composites. This is an emerging technology where the filament contains fibers whose long axis is aligned with the filament, illustrated in Figure 4. This technology has the potential to significantly improve the performance of parts made using polymer extrusion. The desire to exploit these higher properties will bring with it the need to detect discontinuities.

In common with all other AM processes, cracks and porosity are of concern, although the presence of the fiber reinforcement will tend to cause the cracks to be oriented between layers (Z-direction) and adjacent passes (Y-direction), as shown in Figure 5. Contamination of the filament can also result in porosity. Another type of discontinuity seen in ME is lack of fusion, or where the material does not fuse to either the previous layer (Z-direction) or an adjacent pass (Y-direction). The difference between a crack and a lack of fusion is that a crack is due to post-deposit fracture of material that was fused, while lack of fusion means the material was never fused in the first place. In the Y-direction, that can be due to either the passes being too far apart to touch and fuse, or from there being insufficient heat to fuse them together. Another difference is that edges of a crack will be sharp, while lack of fusion will be rounded. A difference between lack of fusion and porosity is that lack of fusion will be elongated in the build direction, while porosity would be equiaxed. Further blurring the definition between lack of fusion and porosity occurs when the distance between passes is barely enough to fill the gap. In this case, the lack of fusion would be a series of short regions in a line, which is much like intermittent porosity.

A final type of discontinuity that is unique to fiber or particulate-reinforced polymer AM processes would be a lack or surplus of reinforcement, or in the case of fiber-reinforced, off-orientation. That is because, unlike fiber-matrix composites, the length of the fibers is not in meters, but closer to millimeters. This means that when the filament is made, there is a probability that a local region would either have a significantly lower or higher amount of fibers than specified, or they could be misoriented. The result would be a local region with greatly reduced or increased strength and stiffness, or the potential for cracking. While this should generally be addressed in the production of the filament, it may be necessary to verify the lack of anomalies during the build phase.

FIGURE 4. Material extrusion (ME) and fiber-reinforced filament.

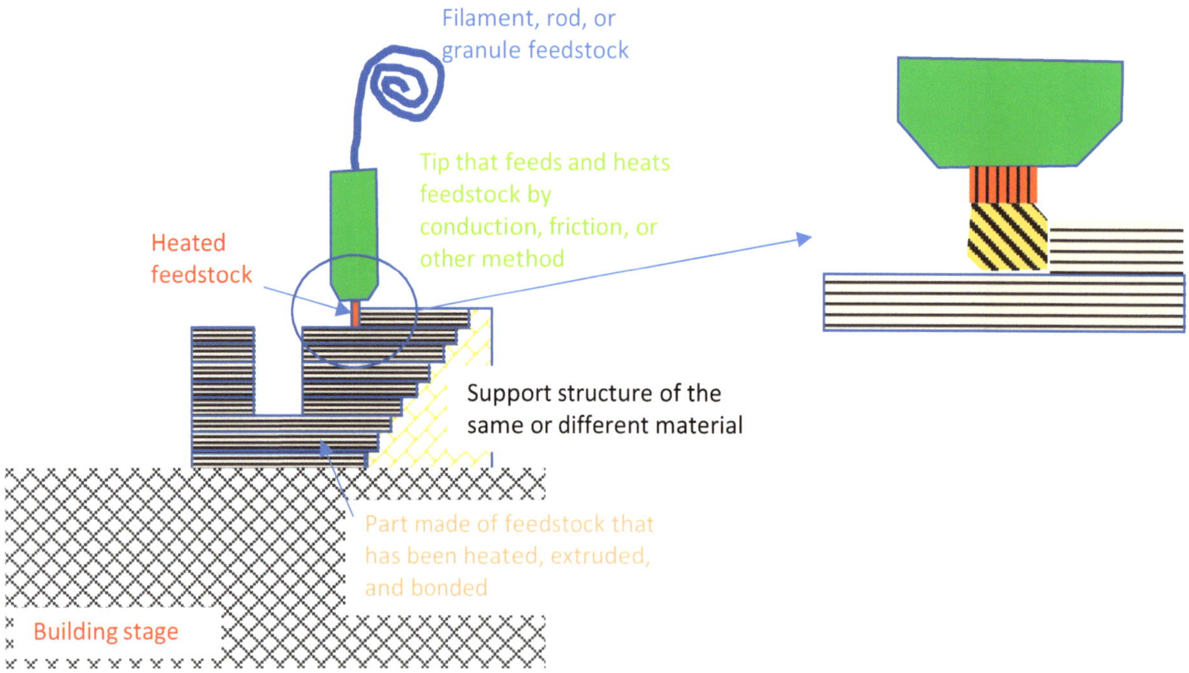

FIGURE 5. Cracks and lack of fusion.

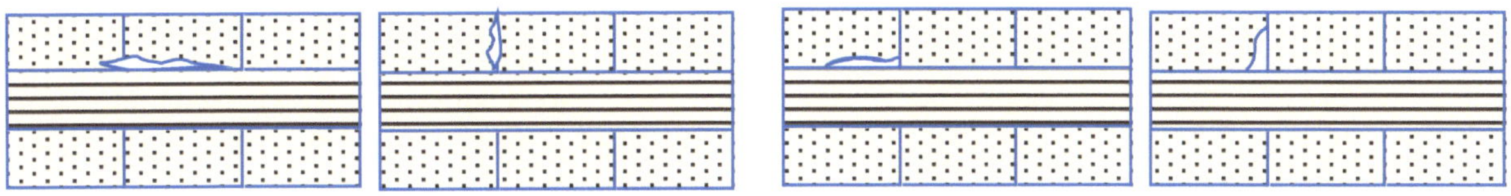

Vat Photopolymerization and Powder Bed Fusion Like ME, polymer composites can also present regions with a lack or surplus of reinforcement. Both neat (polymer only) and composite (particulate or fiber-reinforced) parts made using VP or powder bed fusion (PBF, Figure 6) can contain cracks, pores, and lack of fusion. Pores in PBF can be the result of a number of process conditions:

- **Keyholing**—An excess of energy can either superheat the gas in the powder bed or boil off of the molten powder so that it ejects the molten material.
- **Gas**—The atmosphere can be trapped in part of the build, or excessive moisture in the powder can vaporize. Note that too much vapor will result in keyholing.
- **Shrinkage**—With the exception of water and a few other elements, as molten material solidifies, it shrinks. Normally this shrinkage is fed by the top of the melt pool. If this does not happen, however, the shrinkage will result in a pore.

Thus, it can be concluded that there is a range between too much energy—which results in keyholing—or too little

FIGURE 6. Powder bed fusion (PBF).

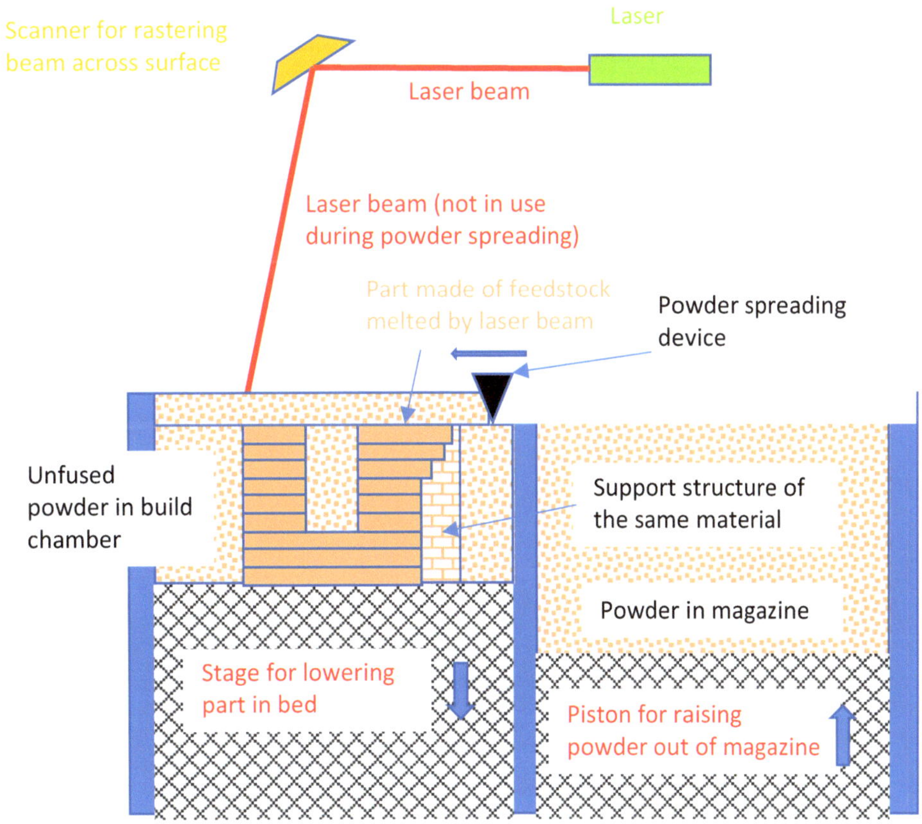

energy, which results in lack of fusion. The wider that range, the more robust the process.

Metals

Powder Bed Fusion Like those made from polymers, metallic parts made using PBF can also have cracks, porosity, and lack of fusion, with some additional causes. Since greater energy is used in PBF, it is possible for melt spatter to cover a region that has not been melted. The extra material prevents full melting and fusing to the previous layer. The frequent use of contour passes (outlining the periphery of the layer either before or after melting the core) in PBF processes provides distinct locations for porosity if not controlled properly (Figure 7). This could be in the form of a keyhole where contour passes overlap with the hatch region too much—resulting in overheating—or the other extreme, which results in insufficient overlap leading to lack of fusion. While porosity at the outer edge of the contour is often removed during post-process finishing, porosity at the inner edge of the contour is either too deep to be removed or is exposed to the surface during post-process finishing. While it is generally agreed that most porosity is due to processing, porosity in gas-atomized powders may provide an additional source of porosity in solidified parts. While inclusions can exist in ceramic and polymer-based processes, they tend to be of more concern in metallic parts where loading and fatigue cycles tend to be higher. Sources of inclusions can be from

- Initial production of the powder, with both nonmetallic (generally ceramic from the melt facility) or high melting point metals (generally tungsten from the melting electrode)
- Contamination of the powder during handling and while in the AM machine (generally considered the higher risk)

This necessitates the very strict contamination control plans that facilities making and using powders must put in place.

Directed Energy Deposition The discontinuities of concern in directed energy deposition (DED, Figure 8) are the same as ME of polymer composites, without the concerns of lack or surplus of reinforcement, but with additional concerns of inclusions and powder porosity in processes that use powder feedstock. In the case of large puddle DED, where the puddle can be up to 12 mm across, lack of fusion can range in size from small (1 mm, typically at a triple point (Figure 9)), to as large as the puddle diameter, with lengths potentially extending several centimeters.

Material Extrusion and Binder Jet Printing The discontinuities of concern in metallic ME and BJP are the same as for ceramic BJP, as both processes use a temporary binder that is removed prior to, or during sintering. Since HIP is commonly performed on sintered metal parts to fully densify them, that means that the expected size of porosity will be lower than in ceramics. It also means that porosity that contains an inert gas may not necessarily be closed and healed, thus a recommendation for the use of vacuum or a reducing atmosphere.

Sheet Lamination Our final process, sheet lamination (SL, Figure 10), especially the ultrasonic version, is mainly concerned with lack of fusion, either between layers in the Z-direction, or between adjacent passes in the Y-direction.

FIGURE 7. Types of Porosity in Powder Bed Fusion.

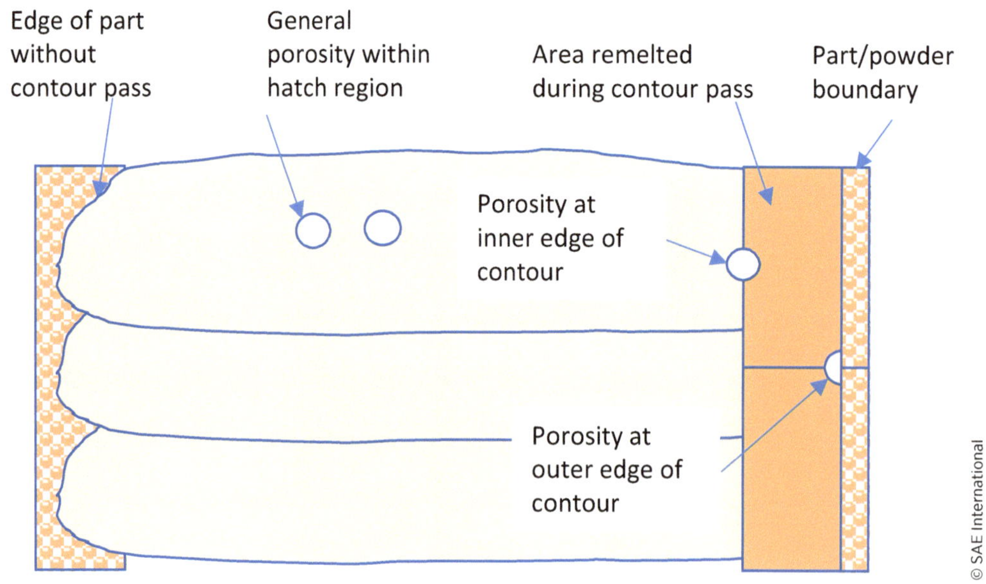

FIGURE 8. Directed energy deposition (DED).

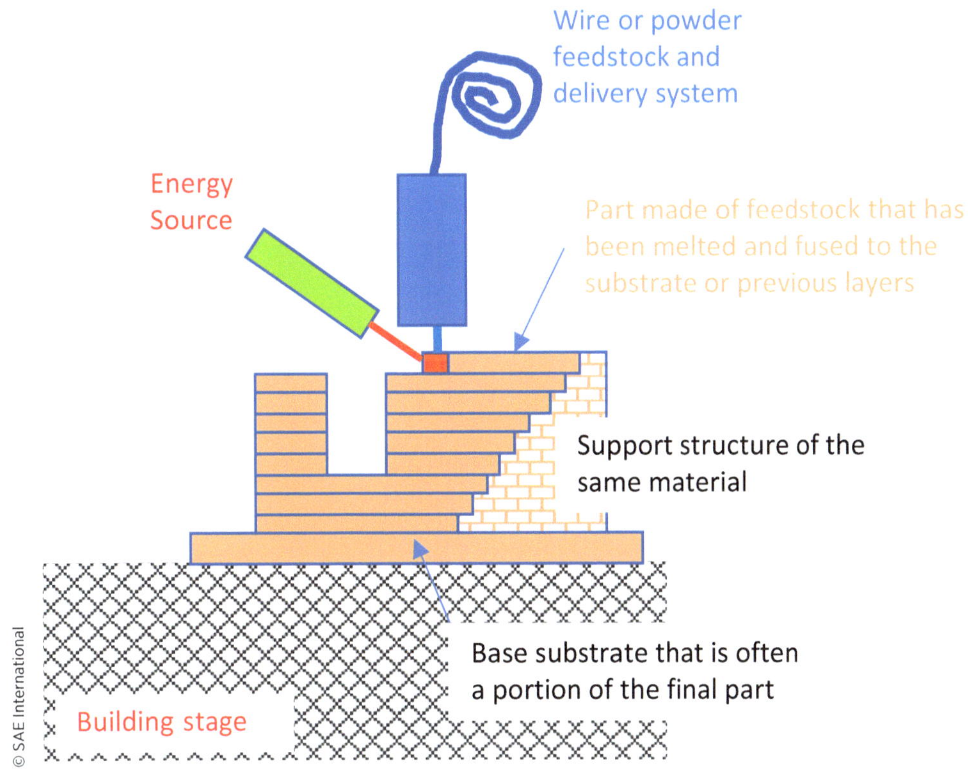

FIGURE 9. Lack of fusion in DED.

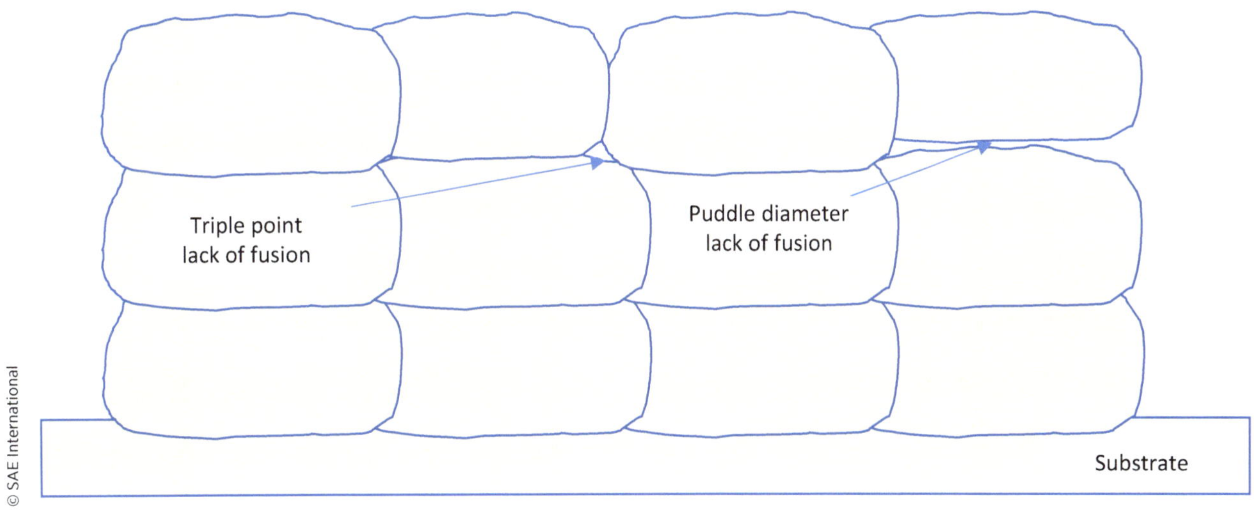

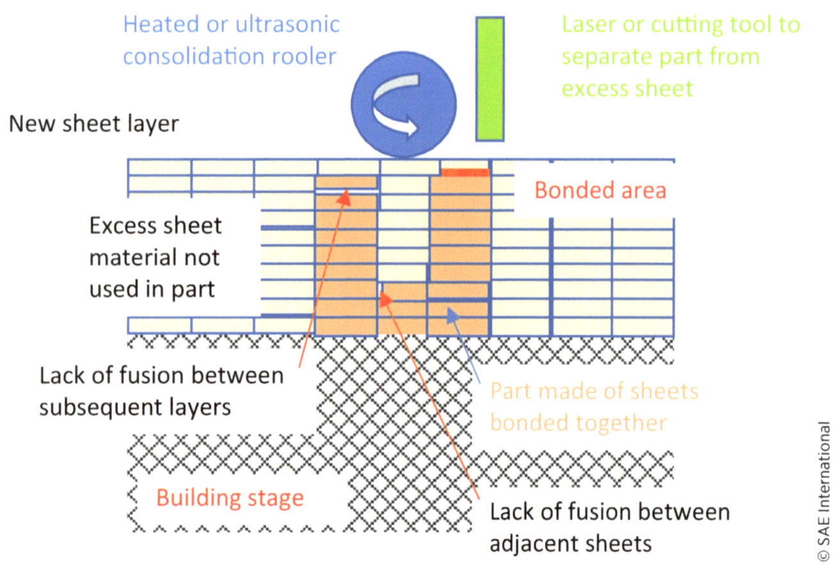

FIGURE 10. Sheet lamination (SL) with lack of fusion locations.

Recommendations

A review of the Table 1 shows that regardless of the material and AM process, there are five types of discontinuities of concern:

- Cracks
- Porosity
- Lack of fusion
- Lack of reinforcement
- Inclusions

The differences between the different material and process combinations are

- The likelihood of a discontinuity occurring
- Its orientation, as some processes (e.g., ME and DED) can build parts and add features in a myriad of different absolute directions
- The potential size relative to that which can be accepted, with most discontinuities being smaller than the layer thickness, with the exception of cracks (which can extend between multiple layers), and lack of fusion, which are most prevalent in the XY plane
- How it occurs (e.g., isolated, in-line, clusters)
- How it would be classified (e.g., dimensions, number in a region, total percent, etc.)

The challenge with a relatively new technology such as AM is determining when to specify NDT. For a product team considering the use of AM and the need for NDT, the following questions need to be addressed:

1. What are the consequences of a premature failure? Loss of a system, loss of life, costly maintenance and downtime, or recalling 1,000,000 vehicles are all much more severe than a poor customer experience.
2. What does my past and present process development and testing tell me are the types, sizes, orientations, locations, and frequency of discontinuities? Understanding the process/material/geometry combination and gathering what may be considered excess data during process development can pay dividends.
3. What does my analysis tell me in terms of a discontinuity compromising the static strength of the part?
4. What does my analysis tell me in terms of a discontinuity growing in fatigue until it compromises the static strength or causes fracture of the part? In the case of AM, discontinuities can often have less of an impact on fatigue life than a rough as-built surface.
5. Is there an NDT method that will reliably find what I am looking for and where am I looking for it? Some parts may only be strength critical or fatigue critical in limited regions.

For a company or an industry in general the following are recommended:

- Understanding the types, sizes, orientations, locations, and frequency of discontinuities for a given process for both in-control, and out-of-control processes to set the boundaries on what can occur
- Understanding the effect of discontinuities on fatigue properties
- Understanding the ability to detect discontinuities that are large enough to be defects

To answer these questions, an understanding of current and emerging NDT technologies and their applicability to AM is needed.

TABLE 1. Summary of discontinuities.

Material System	F42 Process Category	Discontinuity & Orientation
Ceramic	Binder Jet Printing Material Jetting Vat Photopolymerization	• Cracks—Random • Large Pores—No orientation • Undesirable Phases—Random
Polymeric Composites	Material Extrusion	• Cracks—Interlayer (Z) • Cracks - Adjacent Passes (Y) • Lack of Fusion—Interlayer (Z) • Lack of Fusion—Adjacent Passes (Y) • Pores—No orientation • Lack or Surplus of Reinforcement—Random
	Vat Photopolymerization Powder Bed Fusion	• Cracks—Random • Lack of Fusion—Interlayer (Z) • Pores—No orientation, although they can be in-line, clustered, or layered • Lack or Surplus of Reinforcement—Random
Metals	Powder Bed Fusion	• Cracks—Random • Lack of Fusion—Interlayer (Z) • Pores—No orientation, although they can be in-line, clustered, or layered • Inclusions—No orientation
	Direct Energy Deposition	• Cracks—Random • Lack of Fusion—Interlayer (Z) • Lack of Fusion—Adjacent Passes (Y) • Pores—No orientation, although they can be in-line, clustered, or layered • Inclusions—No orientation
	Binder Jet Printing	• Cracks—Random • Pores or Large Pores—No orientation • Undesirable Phases—Random
	Sheet Lamination	• Lack of Bonding—Interlayer (Z) • Lack of Bonding—Adjacent Passes (Y)

Nondestructive Testing Needs, Methods, and Future Developments

Ceramics

As noted in Table 1, the primary discontinuities of concern in this category of AM are cracks and pores that are larger than those that are expected to remain after the sintering process. In parts that are fully or near fully densified, the same discontinuities are of concern, with the allowable size being smaller. While there are always exceptions, especially with fully dense ceramics or materials formulated for unique properties, the physical nature of ceramics strongly compromises the ability of most widespread NDT technologies:

- **Ultrasonic Inspection**—Porosity attenuates sound waves almost completely.
- **Penetrant**—Surface porosity absorbing large amounts of penetrant; discontinuities need to be tight, so capillary action retains penetrant after washing.
- **Thermography**—Low thermal conductivity and porosity greatly limit depth of inspection without exposing the part to high and potentially damaging heat fluxes.
- **Eddy current**—Insulative properties prevent generation of eddy currents.
- **Magnetic particle**—Most ceramics, polymers, and metals have little, if any, magnetic properties, with the exception of ferrous ones.

Therefore, post-build NDT is limited to radiographic and resonance methods, with the potential for process monitoring to detect inhomogeneities during the build.

Radiographic Methods Radiography is the primary method used to perform volumetric inspection of ceramic parts. It works by passing x-rays (gamma rays or neutrons in some cases) through the part and onto some kind of detector. Discontinuities are found because areas with cracks or pores have less matter to absorb or scatter the x-rays. Higher-density inclusions absorb more x-rays while lower-density inclusions absorb fewer x-rays. Thus, radiography is a comparative method of finding discontinuities. Complicating radiographic inspection are changes in geometry, which change the number of x-rays hitting the detector—another example of how in AM complexity increases the difficulty of NDT. A further geometric complication is noise, especially around the edges. Smaller features and their quantity increase the number of edges and the noise associated with them. Radiography consists of a number of different technologies:

- **Film Radiography (FR)** is the original method in which x-rays that pass through the part being inspected expose a piece of film. The more x-rays that pass through the film, the darker the image. A typical sensitivity is the ability to detect a spherical pore 2% of the part thickness.
- **Computed Radiography (CR)** replaces film with a reusable detector plate that is read after exposure to create the image. Benefits of CR over film are digital storage and analysis, larger dynamic range (which means less sensitivity to changes in thicknesses), and generally shorter exposure times. While CR has a larger theoretical resolution size (pixel size being larger than an atom), modern systems have very small pixel sizes, so this is rarely an issue. The prime disadvantage versus FR is the inability to wrap the CR plate around parts or inside of them to achieve the best sensitivity.
- **Digital Radiography (DR)** uses an electronic camera to capture the x-rays coming through and records and stores the image directly. The primary advantage is the ability to manipulate the part during exposure to see different views. A good technique will make use of magnification to achieve the desired level of resolution. Expansions of DR (and sometimes CR) are known as three-dimensional (3D) x-ray and real-time x-ray. These entail taking shots from multiple angles of the same part. The different views can be seen individually, simultaneously, or as a video (real-time) to give a better 3D understanding of the part and any potential discontinuities.
- **Computed Tomography (CT)** is the industrial equivalent of computerized axial tomography (CAT) scanning (Figure 11) used in the medical field, where x-rays that pass through the part are collected by a detector in a series of shots as the detector rotates around the patient. The images are then processed to generate a 3D reconstruction of the part, which would include physical discontinuities. Industrial CT (Figure 11) differs in that higher energy x-rays can be used, and the part (which won't get motion sickness) is often rotated, instead of rotating the x-ray source and detector. Industrial CT systems come in a range of sizes, sensitivities, and resolutions. Sensitivity is a complex relationship between the volume being imaged, the number of pixels in the detector, and the use of magnification in the inspection. The main benefit of CT over the other methods is the ability to generate full-part reconstructions and enabling views from multiple directions and in much greater detail, leaving little room for missing a key orientation. Also, the use of shots from multiple angles improves the ability to detect thin, planar discontinuities, such as cracks. It can also be used in metrology of complex AM parts, where a 3D model of the part, including internal passages can be reconstructed and compared with the original computer-aided design (CAD) file. The primary disadvantage is cost, both in terms of the equipment purchase, the time is takes to image and reconstruct a part, and the time required for interpretation. Another disadvantage is that it must be possible to rotate the part, or the region of the part being inspected in the CT scanner. Depending on the size of the part and the scanner, this can significantly affect the size of what can be resolved relative to the other methods.

FIGURE 11. **Medical (left) and industrial (right) CT scanning equipment (Courtesy Jesse Garant Metrology).**

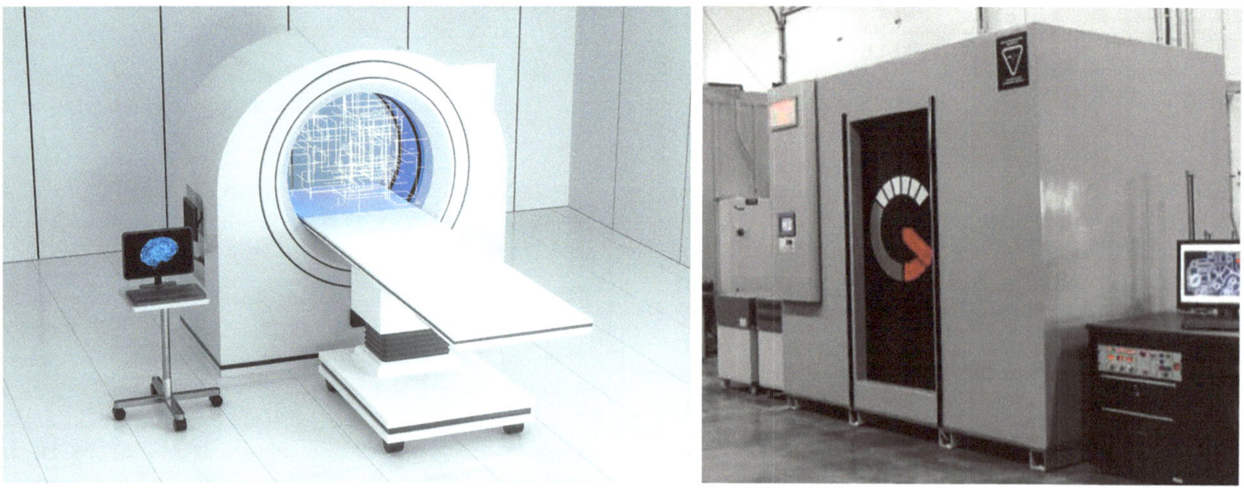

Vink Fan/Shutterstock.com; Reprinted with permission.
Courtesy of Jesse Garant Metrology

Radiography can be quite sensitive at finding pores larger than desired, especially in thinner parts, noting that the general 2% ability for FR refers to the total thickness of material being imaged in a region, so that the pore size for a double-wall part would be 4% of the wall thickness. The low density of most oxide ceramics means that exposure times and voltages will be reasonable. When inspecting porous ceramics, small and homogeneous porosity will appear as a slightly darker gray. Larger, inhomogeneous porosity will create a high level of variation in the background, which will make it harder to detect cracks and unacceptably large pores. Due to the need for the beam direction to be roughly aligned with the orientation of a crack makes radiography less than ideal for detecting individual randomly oriented cracks, although it can be used to find clusters of smaller cracks in a range of orientations. Selection of the desired radiographic method involves a series of trade-offs in terms of desired size detection, cost, and the ability to turn around the inspection.

Resonance Methods The best example of resonance inspection is what happens when one taps a loosely held glass and hears a ringing sound, which indicates its resonant frequency. A glass with a crack or where the wall thickness varies significantly would have a different resonant frequency. Modern resonance inspection can be a precision testing method where the fixturing and method of holding the part and the means of introducing the energy can be highly reproducible. Additionally, resonance inspection is an absolute (versus relative) inspection method where an acceptable frequency range can be specified.

In terms of inspecting ceramic AM parts, resonance methods can be good for identifying those with cracks or excessive amounts of porosity. As the entire part is tested at once, it can be very quick, with the compromise being difficulty of characterizing the type, size, and location of the discontinuity. Thus, it can be used in conjunction with radiography where resonance is used to screen and accept the majority of parts, with radiography used to further investigate and make a final accept/rework/reject determination on the suspect parts. Successful use of this approach requires

- **The ability to set acceptance criteria for resonance inspection to avoid passing parts that have defects, while at the same time avoiding an excess of false-positive indications (calling an acceptable part "discrepant").** An example of this complexity is determining if two small discontinuities are acceptable where a larger one would not be, and relating that to the inspection method.
- **Having a process sufficiently in control where the vast majority of parts are acceptable and will respond consistently to resonant inspection.** One is probably better off performing 100% radiographic inspection than performing 100% resonant inspection and followed up with having to perform radiographic inspection of 90% of the parts. Examples of actors that can result in an inconsistent response would be a wide range of acceptable porosity levels and sizes, or the presence of microcracks.

Finally, as porosity is an effective dampener of sound, thicker parts, and those with high levels of porosity, it will be limited in the ability to be resonance inspected.

In Situ Inspection While ceramic AM parts are made in layers that enable imaging during printing, all of the primary manufacturing methods entail post-build binder removal, and a sintering operation that significantly changes the density and dimensions of the part. Both of the post-processing operations also provide opportunities for the generation of physical discontinuities, such as cracking or insufficient densification in a local area. Thus, the opportunities for in situ NDT are very limited. That said, opportunities for process monitoring do exist. Examples are the use of optical imaging using the appropriate wavelength to identify

- Inhomogeneities in spreading the powder in BJP
- Errors in binder application in BJP, or jetting the powder/binder mix in MJ
- Bubbles or areas of high or low ceramic density in VP

These can be used to ensure that only green parts made within an acceptable range move onto the next stage of processing, which not only lowers the overall NDT cost by keeping parts that would fail NDT from being built, but also reduces the variation in the acceptable parts to allow for better inspection of the parts that are completed.

Recommendations In the case of ceramic AM parts, the primary unresolved issue is determining the need to perform NDT. This is because many ceramic parts see little or no mechanical loads, and a certain level of porosity is often desirable where ceramics are used as a thermal barrier. Even though the nature of ceramics limits the application of NDT methods, a product team implementing AM for ceramic components where physical discontinuities above certain ranges are not acceptable can use NDT in conjunction with general process control and process monitoring to meet product requirements by

- Fully defining the types, quantity, and size ranges of acceptable and unacceptable discontinuities
- Using process control to minimize the conditions to produce them and process monitoring to eliminate the green and brown parts with a high likelihood of producing unacceptable parts
- Targeting NDT for the critical regions of the parts

Polymeric Composites

As noted in Table 1, the primary discontinuities of concern in AM of polymeric composites are cracks, lack of fusion, pores, and inhomogeneous or misoriented reinforcement. Over the past 50 years, the development of polymer matrix composites for primary structures in aircraft has been accompanied by the development of increasingly capable methods for NDT of these composites. The majority of these use ultrasonic inspection, while some structures, such as those containing honeycomb, receive radiographic, thermographic, or stress-optical inspection. The general lack of electrical conductivity and magnetism prevent use of eddy current and magnetic particle methods. On the other hand, the relative insensitivity of polymer matrix composites to surface porosity means that penetrant inspection is not performed, aside from some limited edge inspections using volatile compounds. Therefore, post-build NDT for polymeric composites made using AM follow these same methods. Recall that in this SAE EDGE Research Report that the composites made using AM have short fiber reinforcement, versus long fibers used for aircraft wings.

Radiographic Methods Like ceramic parts, radiographic methods are quite capable of finding pores above a certain threshold, with the general lack of porosity in polymeric composites making detection and interpretation easier. Also, like ceramic parts, radiographic inspection is less than ideal for finding individual randomly oriented cracks, although it can be used in traditional laminates to find clusters of smaller cracks in a range of orientations. Radiographic inspection is able to validate the geometry of internal cavities in AM parts from things as simple as verifying support removal to creating a model of the cavity to compare with CAD files. The low density of polymer matrix composites makes radiographic inspection easier. A final consideration in the use of radiographic methods is that the greater geometric complexity and shorter fibers in polymeric composites made using AM mean that there are more orientations for lack of fusion to occur, which complicates radiographic inspection.

Ultrasonic Methods While ultrasonic inspection is the dominant way of performing NDT for large laminates found in aircraft structures, the greater geometric complexity requires a greater level of sophistication for polymer composites made using AM. The primary methods of performing ultrasonic inspection, shown in Figure 12, can be broken into pulse-echo and through-transmission types, and use frequencies between 1 and 10 MHz, well above the maximum audible frequency of 20kHz. Almost all ultrasonic inspection methods use a liquid medium (called a couplant) to couple the sound from the transducer into the part, and back again. Laser-based methods, which use pulsing lasers to create the sound on the part surface, and to act as an interferometer to measure the reflected signal are beginning to see use.

- **Volumetric pulse-echo inspection methods** work much like sonar does on a ship. A pulse is sent out from a transducer (black square in Figure 12) that

FIGURE 12. Ultrasonic inspection methods.

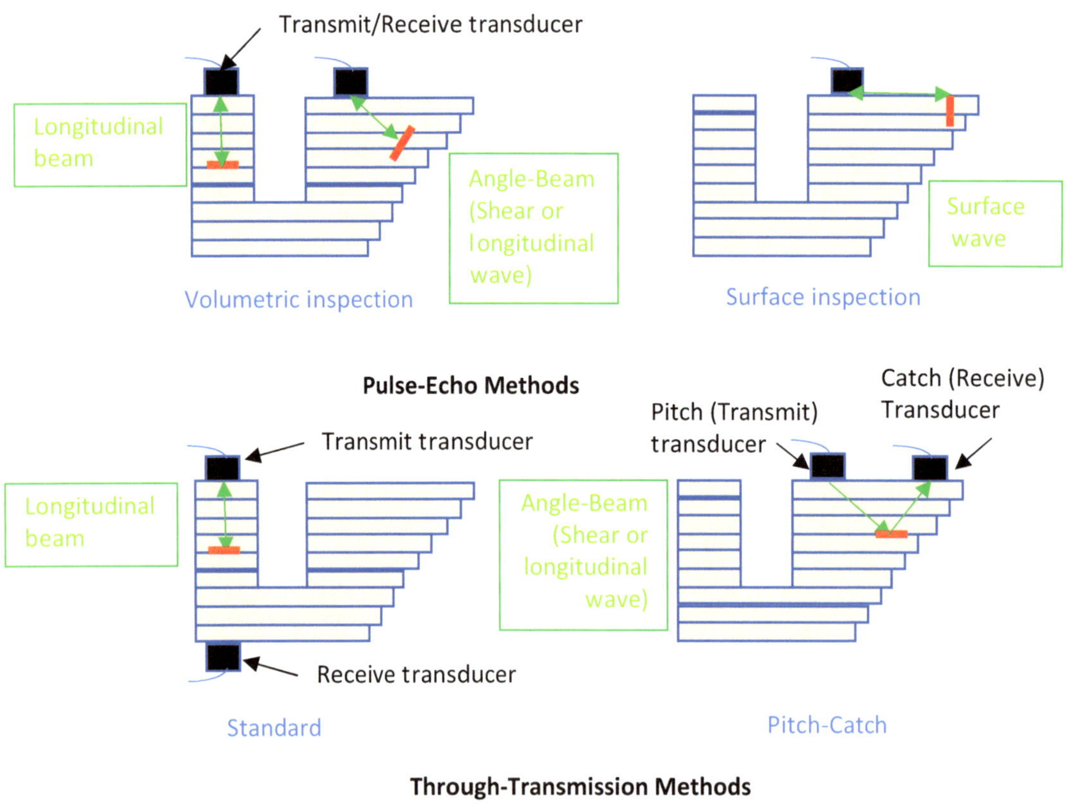

will reflect off of an internal discontinuity (red line in Figure 12). The sound will reflect off of the discontinuity and back to the transducer, where the amplitude of the signal and the time from when the pulse is originally transmitted is noted. The time provides information on the distance of the discontinuity from the surface, while the amplitude (sound volume, measured in dB), combined with the time, provides information on the size of the discontinuity. While the preferred orientation of the beam is to be perpendicular to the axis of the discontinuity, waves can also be reflected back off of the curved surfaces found in porosity—although the amplitude will be significantly reduced. Longitudinal waves are introduced perpendicular to the part surface, while longitudinal or shear waves can be introduced at angles to the surface. In cases where the surface is not flat, but still predictable (such as a constant radius), special adapters, called shoes, are used to introduce the sound into the part. More sophisticated methods will use a phased array of transducers to simultaneously cover a wide range of angles. Typical minimum detection sizes in polymer matrix composites at 5 MHz are 3 mm by 3 mm.

- **Surface pulse-echo inspection methods** introduce a shallow wave into the part that races along the surface and will reflect back off of a discontinuity at or near the surface. This has been rarely used in standard polymer matrix composites, for the same reason that penetrant inspection has not been performed. This could see more usage in AM parts, although geometric features, such as surface protrusions and the roughness between layers, will tend to either contribute to high background noise or false-positive indications.
- **Standard through-transmission methods** introduce a longitudinal beam into the material with one transducer, while listening for the amplitude of the signal with the other. This is the preferred method of detecting small porosity in composites, as the pores will attenuate the amplitude of the signal, which can be correlated with the level of porosity in the composite. While this is widely used for standard composites, one limitation is the ability to access both sides of the part.
- **Pitch-catch through-transmission methods** entail introducing an angled beam into the part with one transducer, while listening for a reflection (or sometimes the lack of reflection) with another transducer. The use of two transducers can accommodate unique part or discontinuity geometries. Pitch-catch is sometimes also used for surface inspections.

An example of where complexity increases the difficulty of NDT is that the following are needed to perform ultrasonic inspection:

- One or more surfaces where the ultrasound can be introduced into the part. Unlike radiography, where a rough or irregular surface may provide background noise, a rough surface can actually prevent ultrasonic inspection from being performed. This includes the undercuts and notches between layers, as well as the rough surface on the bottom and top.
- The part being inspected needs to be able to carry sound. Highly porous parts (note ceramics) will completely attenuate the sound in a matter of micrometers, preventing the transmission or reflection of a signal to analyze.
- In the case of pulse-echo methods, there needs to be sufficient distance for the beam to travel to avoid near-field effects. The methods used for bulk wrought metal parts (e.g., bar, plate, etc.) have near and far fields of 5 mm or so, hence why many metal specifications do not call out ultrasonic inspection for thicknesses under 12 mm. The use of higher frequencies and smaller transducers can decrease the near and far fields, with 0.5 mm commonly achieved in polymer matrix composites. This is accompanied with the trade-off of increased inspection time, more sensitivity to surface finish, and reduced depth of penetration.

Because the discontinuity of interest must have sufficient surface area perpendicular to the beam to either reflect a signal back or to attenuate a through-transmission signal, there is a complex relationship between the surface where the sound is introduced into the part and the orientation of discontinuity of interest. Early carbon/epoxy laminates were flat, and the discontinuities of interest were either planar and parallel to the surface or a porosity. This made ultrasonic inspection rather straightforward. As co-cured structures with stiffening features were developed, it took over a decade to develop a means to inspect even these relatively simple geometries, and to demonstrate their effectiveness. Complex AM part geometries mean that planning and programming the scan pattern can become very complex and time-consuming, while demonstrating effectiveness is more challenging, with there still being the possibility of significant no-inspect areas. Thus, identification of the critical areas of a part and designing to accommodate ultrasonic inspection are more necessary.

Thermographic and Stress-Optical Methods The final two methods are image-based and are most commonly encountered in the NDT of bonded honeycomb or foam structures. Thermographic inspection involves applying a transient thermal load to a part and viewing it with an infrared camera—in short, looking for hot or cold spots which indicate reduced thermal conduction for a region that is not bonded or poorly bonded. Stress-optical methods involve viewing a part with a special camera that looks for changes in an interference pattern before and after applying a low stress (often with a vacuum). This change in the interference pattern shows regions where the local stiffness is anomalous, which would be the result of not being bonded or even weakly bonded (known as kissing bonds). Because both methods view relatively large areas with a camera, they can be quite rapid for inspecting large parts. Discontinuity (lack of fusion) sizes tend to be significantly larger than what can be found using radiographic or ultrasonic methods, keeping in mind that bonded honeycomb or foam assemblies—by definition—contain physical discontinuities (e.g., cell size, pore size). AM parts made using polymer composites that meet these characteristics would be good candidates for NDT using these methods.

In Situ Inspection The simple reason that inhomogeneous and misoriented reinforcement have not been discussed so far is because it would be exceedingly difficult to detect with the above technologies. Because the difference in density between the polymer matrix and a carbon, glass, organic, or oxide fiber are quite small, a deviation would have to be quite abrupt to detect with radiography. The relatively small size of the fibers and a subtle change in content would also be difficult to detect using ultrasonic inspection, as there would be little or no interface to reflect off, and the difference in attenuation for amplitude-based methods would be small as well. Since fiber content has a significant impact on stiffness, the right kind of stress-optical method could work.

As a post-build NDT method—the fourth method—thermographic is more than likely infeasible, as small changes in fiber content would not have a dramatic impact on thermal conductivity of a part with many layers. As an in situ NDT method, however, thermography or another type of radiant inspection method has potential. This is because a change in the fiber volume within a single layer could have a significant effect on the thermal conductivity of that single layer. Also, since polymers vary in their absorption and reflection of different parts of the electromagnetic spectra, it can be envisioned that an imaging method sensitive to changes in reinforcement could be developed. The same type of electromagnetic methods could be applied in the detection of cracks, pores, and lack of fusion, and detection would generally be easier. The primary challenge would be developing and calibrating the filters and thresholds to obtain the desired balance between false positives (indication of a discontinuity below a size of concern) and false negatives (missing a discontinuity above a size of concern).

Finally, the use of in situ ultrasonic pulse-echo methods may need to be considered. Laser ultrasonic technology has progressed to the point where systems are now available, although they are not an option for current industrial polymer AM machines.

Recommendations Like ceramics, the primary unresolved issue regarding the NDT of polymeric composites made through AM is determining the need to perform it in the first place. As the AM of polymeric composites moves to more highly loaded and structurally significant applications, the answer will increasingly be yes. Therefore, in addition to the recommendations made for ceramic parts made using AM, a product team implementing AM for polymeric composites needs to consider the following:

- Designing the part so that conventional post-build ultrasonic inspection methods can be used in the highly-loaded areas. This is primarily a "design for AM" (DfAM) consideration.
- The use of in situ inspection methods for volumetric inspection where conventional ultrasonic inspection is not possible, with a goal to eventually making any post-build ultrasonic inspection unnecessary. Note that this may require development and validation of a new technology.
- The use of a method to inspect for surface cracks, especially edges, and the development of one if no current method exists.

Metals

Metals—the primary material in the mobility industry over the last 200 years—still dominate structural and mission-critical part application. As such, the majority of NDT processes were originally developed for them. While the types of discontinuities of concern are similar to those in other material systems, there are differences in the NDT aspects of metals from other systems. These differences can be broken into the categories of material thickness, discontinuity size, and location.

The first of these—material thickness—is mainly a comparison between wrought metals and polymer matrix composites. While it is rare to see polymer matrix composites thicker than 3 cm, metallic components are commonly thicker than 10 cm, especially in load introduction areas. This, combined with the higher density of metals, presents a limitation in performing radiographic inspections. The advent of topological designs for AM parts actually reduces this difference in thicknesses, which will help with radiographic inspections, especially in simplifying radiographic inspection and thereby reducing the cost.

The second difference—discontinuity size—is primarily related to ultrasonic inspection. The greater homogeneity of and the reduced sound attenuation in metals means that discontinuities as small as 400 μm diameter can be detected in some metals, with 1.2 mm diameter a common criteria, and 3 mm for the thickest parts; while detection requirements for polymer matrix composites are rarely smaller than 3 mm diameter, with 10 mm diameter more common.

Third is the need for surface NDT. Besides relatively simple examinations for pores and cracks, specialized surface NDT methods are rare in ceramics and polymer composites.

In the case of metals, however, the surface is often the region of most concern, hence the development of a wide range of surface inspection methods.

Surface Inspection Methods and Geometric Complexity The final three widespread NDT methods—penetrant, magnetic particle, and eddy current—are used for surface or near-surface inspections and are nearly exclusive to metals. As surface or near-surface inspection methods, their usefulness is also highly dependent on the smoothness of the surface.

Penetrant inspection can only detect discontinuities that are open to the surface. This is because the method involves putting a visible or fluorescent liquid on the surface of the part and allowing capillary action to draw the penetrant into cracks or fissures on the surface. The bulk penetrant on the surface is then removed, and a developer is applied, which draws the penetrant in the cracks of fissures back to the surface (Figure 13).

While detection capabilities for linear discontinuities using high-sensitivity fluorescent penetrant is 1.5 mm, several factors can significantly compromise this ability:

- Visible penetrants are less sensitive than fluorescent ones.
- The penetrant will wash out of shallow pores, preventing them from being detected.
- Greases, oils, and other contamination will prevent the penetrant from entering the discontinuity.
- Machining of surfaces on metal parts with yield strengths under 1500 MPa can smear metal across the discontinuity, preventing the penetrant from entering. This is avoided by use of a light etch to remove approximately 10 μm from the surface, opening up the discontinuities. Note the cleaning and etching processes tend to also remove other contaminants. Some surface improvement processes (e.g., bead blasting, etc.) can smear metal as well.
- Surface roughness is the primary compromising factor regarding AM. The undercuts and notches between layers on the surface of AM parts trap penetrant, which is then drawn to the surface of the part by the developer, resulting in false-positive indications that overshadow any real ones.

FIGURE 13. **Visible penetrant inspection of a weld showing a crack.**

High Simple/Shutterstock.com

- Surface porosity in sintered parts made using binder jetting will also trap and re-release penetrant, although a potential solution is to chemically etch the surface to remove the layer containing surface pores. Note that this would also require the use of HIP of the parts to heal the internal pores just below the surface to prevent the etch from exposing them.

Unlike penetrant, magnetic particle inspection can detect discontinuities that are not open to the surface, although the subsurface depths greater than 1 mm would be uncommon. This process involves inducing a magnetic field in the part, and then applying magnetic particulate matter to highlight areas on the surface where the magnetic field is disturbed. Figure 14 shows the inspection of a pipeline weld.

One benefit of magnetic particle over penetrant is that no pre-penetrant etch is required, and similar detection capabilities are obtained. Drawbacks to magnetic particle inspection are

- The material to be inspected needs to be ferromagnetic.
- The process is more geometry dependent, requires a part-specific procedure, and generally greater operator skill than penetrant.
- Like penetrant, the undercuts and notches between layers on the surface of AM parts will disturb the magnetic field—the result is a part with so many false-positive indications that detection of anything, other than large cracks cannot be done.

Like magnetic particle inspection, eddy current inspection can detect discontinuities that are not open to the surface, although depths greater than 1 mm on a general surface inspection would be uncommon. This process uses an electromagnetic coil to induce eddy currents in the part. Physical discontinuities in the part will disturb the eddy currents and provide a response in the test instrument. The electromagnetic coil is contained in a probe, of which there are two primary types:

- Surface probes are like a pen where the operator scans across the expected discontinuity.
- Bolt hole probes rotate inside a hole, which is a common location for service-induced cracks.

Surface and bolt hole probes can be combined in large scanning devices as well. One benefit of eddy current is that since it does not require liquid couplant like ultrasonic inspection, hazardous x-rays like radiography, liquid penetrant, or particulate matter, it is often the preferred method for performing field inspections, especially inside aircraft structures. The primary disadvantage is that with the exception of some specialized equipment, it is very labor intensive, requiring manually scanning a surface in 1 mm wide swaths. Hence why it is primarily used for targeted inspections of parts. Also, like magnetic particle inspection, the undercuts and notches will disturb the eddy currents, providing false-positive indications. Thus, a smooth surface is also a consideration if a part may need in-service inspections.

While a rough surface significantly compromises the ability of surface inspection methods in metals, it also impacts radiographic and ultrasonic inspections. Surface finish impacts ultrasonic inspection by making it very difficult to couple the sound into the part. Not only does this reduce the strength of the signal to analyze, but the rough surface acts like a bunch of reflective surfaces, increasing the overall noise of the signal. In through-transmission inspections, the rough back surface creates additional noise while further reducing the strength of the transmitted signal. In pulse-echo inspections, the reflection off of the back surface, which is often used to validate that sound has been introduced into the part, is scattered. Rough surfaces can also produce false-positive indications.

Funtay/Shutterstock.com

FIGURE 14. **Magnetic particle inspection of a pipeline weld.**

FIGURE 15. Radiograph of a cylindrical AM part comparing geometric noise in the as-built portion to the machined portion.

Reprinted with permission. Courtesy of Jesse Garant Metrology

In the case of radiographic inspections, the rough surface raises the overall noise level of the radiograph, making it more difficult to resolve the regions near the surface. When combined with the more complicated geometry and thinner sections found in AM parts, this high noise level can be present in a significant volume of the part (Figure 15).

Ultrasonic inspection is the NDT method most impacted by geometric complexity. The vast majority of ultrasonic inspections are currently performed on flat parts that are generally planar, or simple cylindrical parts. The actual geometry of some die forgings, especially for turbine engine discs, is driven by the ability to perform an ultrasonic inspection of the desired sensitivity. Ultrasonic inspection of weldments and castings are generally limited to small regions of simple geometry.

Figure 16 shows some of the challenges in performing ultrasonic inspection of simple stiffeners in a large (10 mm diameter) puddle DED part—which, by and large, have been solved using good DfAM practices. The left stiffener shows how the curved, irregular crown causes the sound to scatter, making inspection ineffective. While the stiffener on the right shows how machining the crown off can enable good sound transmission, the sound beam near the side of the stiffener is reflected. This means that there is a zone along the surface where a good inspection cannot be performed. In the case of preforms made using large puddle DED, this is

FIGURE 16. Ultrasonic inspection challenges in a large puddle DED part.

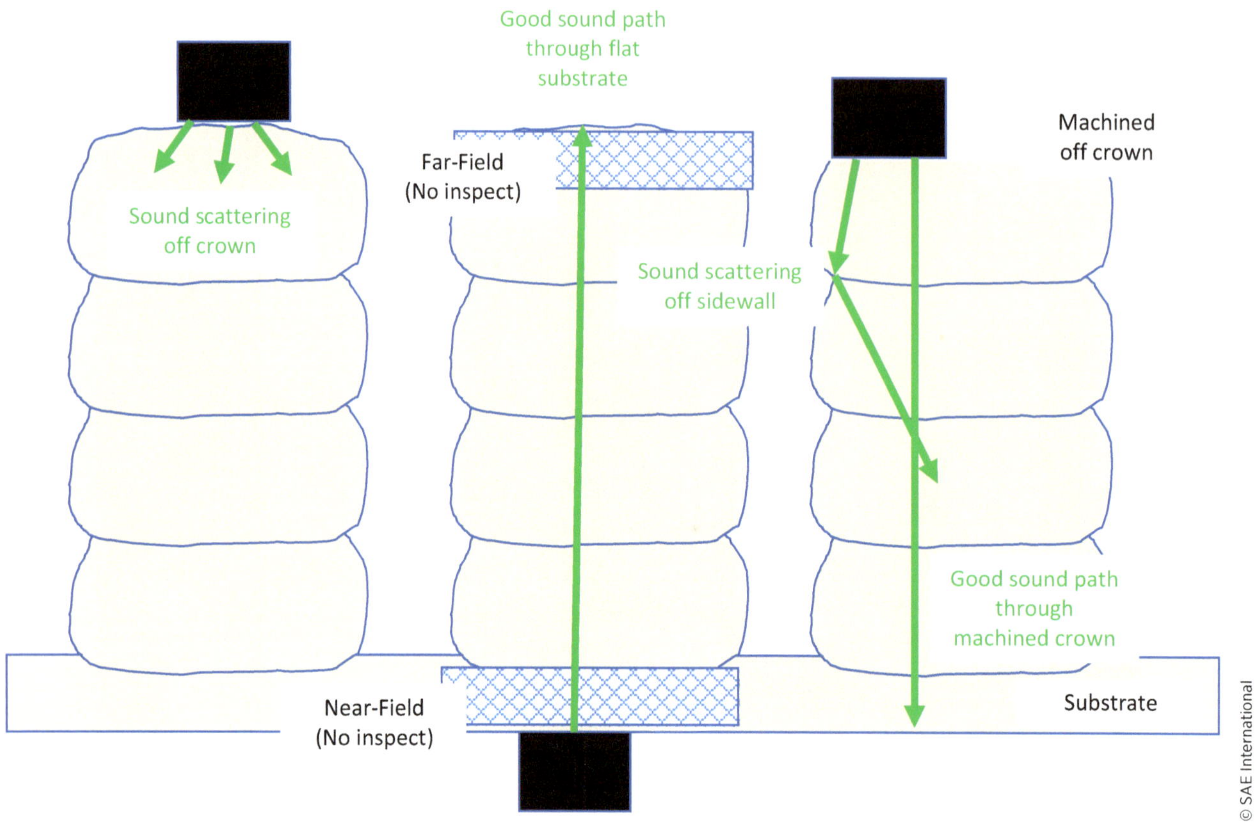

often machined off, so the lack of inspection is not an issue, but it is a design consideration that impacts having a minimal "buy-to-fly" ratio (i.e., the ratio of material bought to make a part versus the weight of the part as installed). The center stiffener shows another option, which is to introduce the sound through the flat substrate. This is also used to illustrate the near-field and far-field zones where a good inspection cannot be obtained, which would apply to all three stiffeners. Also, note that the crown in the center stiffener still needs to reflect back sufficient signal to verify that the inspection is being performed. It can be inferred from this that ultrasonic inspection of complex topological shapes can be almost impossible.

The combination of geometric complexity and surface roughness in metallic AM parts combine to make performing NDT a significant technical challenge if metal AM is to advance to more demanding applications. This applies to the aerospace industry, which is accustomed to 100% volumetric and surface inspection to a high degree of sensitivity. It can also apply to parts of the mobility industry where the high level of process robustness—which AM is still striving to meet—means that NDT is not a standard practice. Having this capability is one of the primary barriers to widespread implementation of AM in these applications. This is because of the impact discontinuities have on the fatigue and fracture of cyclically loaded parts.

Fatigue and Fracture Considerations While all material systems are susceptible to fatigue and fracture, the combination of applications and loading conditions mean that fatigue and fracture considerations often drive the design of metallic components in the mobility industry. Thus, where NDT is required, the allowed discontinuity sizes in metal parts are relatively small, with volumetric and surface inspections being required. The excellent record of wrought products in aerospace, combined with the fact that both volumetric and surface NDT are still required, suggests that a new technology like AM will require NDT in aerospace for a significant period of time. It also indicates that not only are the products generally free of physical discontinuities larger than the acceptance criteria, but that they are free of physical discontinuities of virtually any size. Since it will be even more difficult for metal AM parts to demonstrate a similar level of freedom from physical discontinuities, it may be advisable to assume an acceptable population of small (10-200 μm) discontinuities in metal AM products, and use a probabilistic method to determine fatigue properties. The unresolved issue in this case is determining the NDT technology or technologies to be used and the required sensitivity level. In industries where 100% volumetric and surface NDT is not generally required, but where metal AM does not have an extensive production and service history, the unresolved issue is determining if it needs to be done in the first place, and to what extent.

Future Developments For the trend in metal AM toward more complex geometries and more structurally significant parts to continue, a means to avoid post-build NDT methods with the limits mentioned above must be developed. This means having effective in situ NDT methods that take advantage of the shallow depth and simple geometry of inspection needed when performing it as each layer is built. This can also be aided by better process control that reduces the number of physical discontinuities in the first place. The development of in situ NDT methods for a number of metal AM processes by research organizations and AM equipment builders is already underway (with promising results).

Process control methods being developed and implemented in powder bed processes (BJP and PBF) include imaging each layer after the powder is spread. This finds and enables correction of inhomogeneities in the powder layer before fusion or binding is performed. Other, more complex approaches will combine the outputs of multiple sensors to determine if a process is operating within a window. The sensors could be ones that view the fusing process, along with those the look at the powder bed, and those that monitor other aspects of the machine. Similar approaches are being implemented in DED processes, such as monitoring puddle temperature/radiance, shape, and other features. Because of the way most DED processes deposit and melt the feedstock in a single curvilinear motion, the use of a laser ultrasonic inspection following the melt pool is also possible.

Near-infrared and infrared imaging methods are being developed and implemented that look at differences in how a fused region cools to indicate if there is a physical discontinuity, with some research looking into ways to repair an area with a discontinuity prior to building the next layer. An example of a near-infrared-based in situ method is EOSTATE Exposure OT, which images each layer to identify physical discontinuities and generates a report for interpretation (Figure 17). Systems that monitor cooling of the puddle are also in work for DED processes. Additionally, there is significant potential for further growth in the methods with the advent of multispectral cameras, improved data analysis tools, and machine learning.

While in situ NDT technologies are being developed and implemented, two unresolved issues remain:

- How to validate the capability and reliability of the technologies
- How to ensure that no discontinuities were created after in situ inspection of a layer or layers, especially cracks caused by residual stresses

Extensive studies have been performed for conventional NDT methods on wrought and cast products to develop what is known as the probability of detection (PoD), which indicates the likelihood that a given discontinuity will be detected. In post-build NDT, these are performed by artificially introducing discontinuities into parts covering the geometries of interest and having NDT performed by technicians who are unaware of the presence and location of them in a given part. In the case of in situ NDT, methods to reliably and repeatedly produce said discontinuities will have to be developed, followed by destructive testing to determine the actual characteristics of the discontinuity. It will also be necessary to identify the geometric factors that influence detection and include them into the study. One aspect that will simplify the development and validation of in situ NDT is the ability to perform different analysis methods to determine which method provides the desired detection capability with a minimum of false positives (although this will require full definition and locking down of the validated analysis method).

Resolving the second issue will first require careful study what types of discontinuities can occur after in situ NDT of a layer or layers is complete, along with means to detect it with a post-build inspection. Needless to say, this must be significantly simpler than typical post-build NDT. It can be envisioned that, in the case of cracks from residual stresses, this may entail the combination of a surface inspection with a resonance method.

Recommendations In order for metal AM to expand in terms of industries, complexity, and structural significance, the capability of NDT needs to increase (primary), while the cost needs to decrease (secondary). In addition to the recommendations for ceramic and polymeric composites, for industries where volumetric and surface NDT is expected, the following is recommended:

- Product teams need to incorporate NDT as part of DfAM to ensure that regions of parts that need to receive NDT are designed in such a way that it can be performed effectively.
- Surface inspections need to be made viable, which will require the development of technologies to affordably do so over a complex geometry. Note that in the case of large puddle DED of preforms that are 100% machined, the machining operation accomplishes this.
- Engineering and research organizations need to investigate, develop, and validate in situ NDT methods. Additionally, these methods need to be correlated with the discontinuity sizes of interest and their potential impacts on performance.

For industries where volumetric and/or surface NDT is a special case, one recommendation that can help a product team make this determination takes advantage of the significant reduction in fatigue performance caused by the rough surface [1]. If a product team determines that a part with a rough surface will still meet performance requirements, the size of an internal discontinuity that would compromise performance may be quite large and its location limited to a few critical locations.

FIGURE 17. **EOSTATE Exposure OT In Situ NDT with schematic (top), layer image (lower left), and reconstruction (lower right).**

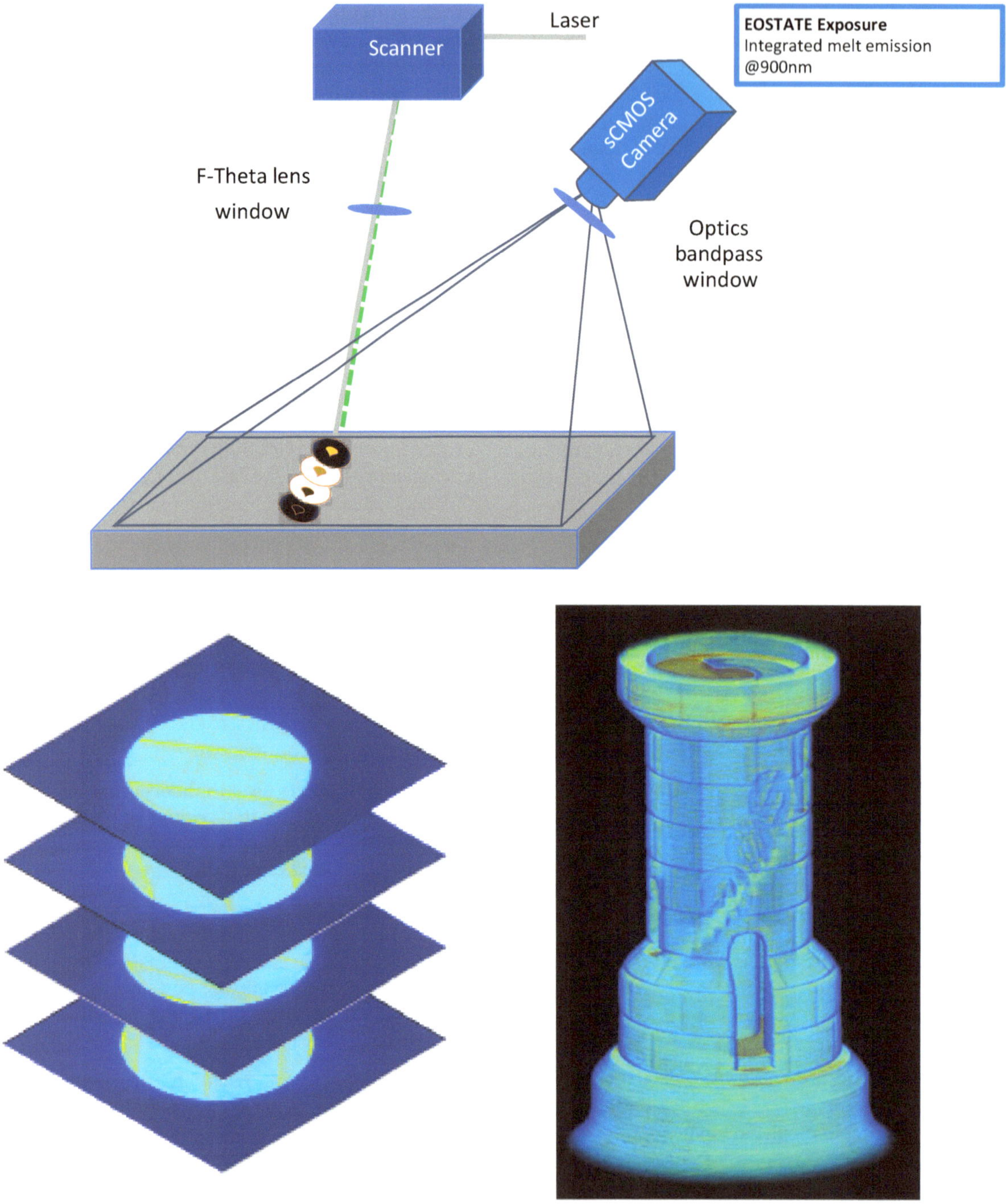

Reprinted with permission. Courtesy of EOS

Summary

Determining the Need to Perform NDT on Additively Manufactured Parts in the Mobility Industry

As AM transitions from prototyping to more demanding applications, the need to assure that a part is free from physical discontinuities will become more important. While processes that are robust, controllable, and capable of being monitored can keep discontinuities from existing in the first place, at some level of criticality it will be necessary to perform NDT to provide greater assurance. In order to determine the need for NDT of an AM part as well as the desired method, a product team must know the following:

- The consequence of a failure on the overall system
- The impact of physical discontinuities on the ability of the part to meet performance requirements
- The cause, size, orientation, and likelihood of potential physical discontinuities
- The available NDT technologies to detect the physical discontinuities of interest with confidence
- How the design of the part and the manufacturing sequence can impact the ability to detect a physical discontinuity of interest
- The cost and ideal position in the manufacturing process to optimize the chances of detection and minimize the loss in value added if a discontinuity is found

These determinations are the same regardless if AM or another manufacturing technology is being considered. What makes this determination more difficult in AM is the geometric complexity, surface roughness, and the current paucity of historical data. Therefore, it is incumbent on the product team—and especially the research and engineering organizations that support them—to begin developing the data needed to make these determinations. Figure 18 provides a guide that a product team can use as a starting point in making this determination, be it process monitoring, NDT, or process monitoring with a low threshold to require post-build NDT. Note that while consequence of failure and impact of discontinuity are combined into a single axis, a product team can treat these as independent axis for their analysis. It should also be noted that as a successful production history on a part or process is built up, that the need for NDT can be decreased. Finally, product teams and the research, engineering, and quality organizations that support them need to do the following:

- Develop methods to build and track production history and correlate it with post-build NDT results.
- Develop and qualify process monitoring methods to reduce and eliminate post-build NDT for the applicable parts.
- Investigate, develop, and qualify in situ NDT methods to eliminate the constraints of post-build NDT on part design.

Having the Ability to Perform NDT on Additively Manufactured Parts in the Mobility Industry

In determining the NDT of AM parts, it is much easier to say that product teams need to have the data than to actually have or develop it. Once a determination has been made that some form of NDT is needed, an organization then needs the ability to reliably perform it. While this has been successfully addressed for some process/application combinations, such as fully machined preforms made using large puddle DED, it is still unsettled in others. The combination of geometric complexity and surface roughness greatly complicates the ability to reliably and cost-effectively perform post-build NDT. Therefore, at some point the successful development and validation of in situ NDT, improved surface-finishing techniques to enable surface inspections, and simple post-build inspections for later-developing physical discontinuities will be required for the mobility industry to fully exploit AM in its pursuit of greater efficiency and safety.

SAE EDGE™ Research Reports

SAE EDGE™ Research Reports, like the present report on "Unsettled Topics on Nondestructive Testing of Additively Manufactured Parts in the Mobility Industry," are intended to push further out into still unsettled areas of technology of interest to the mobility industry. SAE launches these reports before attempting to form a joint working group, let alone a cooperative research program or a standards committee.

SAE EDGE™ Research Reports are intended to be quick, concise overviews of major unsettled areas where vital new technologies are emerging. An unsettled area is characterized more by confusion and controversy than established order. Early practitioners must confront an absence of agreement. Their challenge is often not to seize the high ground but to

FIGURE 18. NDT decision guide.

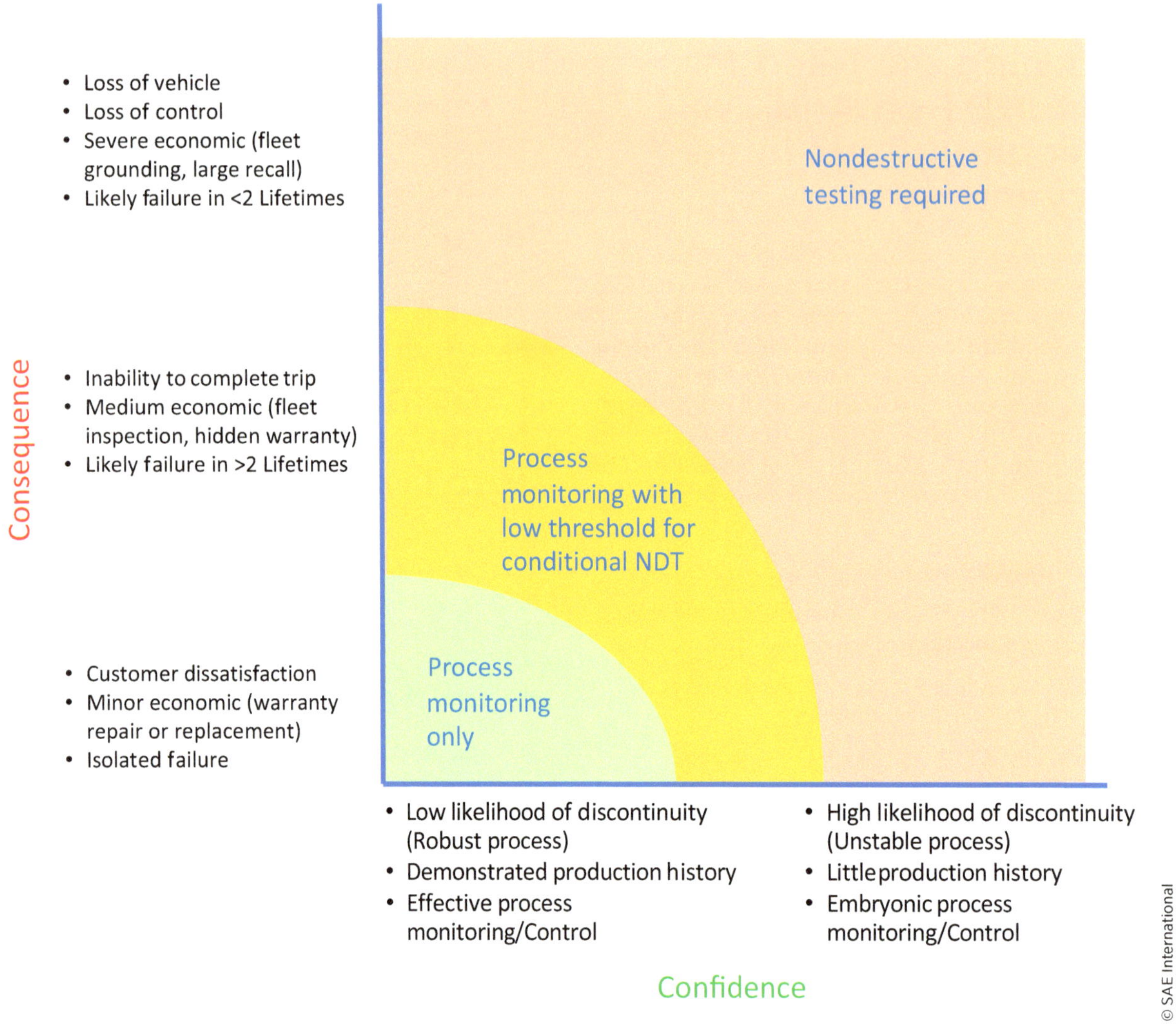

find common ground. These scouting reports from the frontiers of investigation are intended merely to begin the process of sorting through critical issues, contributing to a better understanding of key problems and providing helpful suggestions about possible next steps and avenues of investigation.

SAE EDGE™ Research Reports, therefore, are fundamentally distinct from the more formal working groups approach and far removed from the more mature research program and standard's development process.

Next Steps on NDT of AM in the Mobility Industry

This publication should be considered only as a first step toward clarifying the issues around NDT of AM in the mobility industry. The intention behind this and other SAE EDGE™ Research Reports is to start a dialogue among interested parties on important industry-wide topics that require further attention. The expectation is that these explorations of unsettled areas of technology will lead to the formation of working groups and, ultimately, committees that can address and resolve the issues they raise, producing a framework for developing a common vocabulary of definitions, best practices, protocols, and standards needed to support continued progress toward safer and more innovative automotive products.

The collaboration that gave rise to this publication demonstrated a great willingness on the part of the industry to define the terminology, procedures, and eventually the standards needed to enable NDT of AM in the mobility industry technology to move ahead as quickly and efficiently as possible. SAE International has demonstrated its lead in this and closely related areas by the development and promulgation of specifications for AM.

This SAE EDGE™ Research Report on "Unsettled Topics on Nondestructive Testing of Additively Manufactured Parts in the Mobility Industry" identifies the following key topics for further pursuit, both through continued informal discussions among industry practitioners and through more formal working groups:

- Establish industry- and process-based working groups to develop methodologies and standards for validation of in situ NDT technologies, along with methods for certification of inspectors.
- Establish industry-, process-, and material-based repositories for documenting the occurrence of physical discontinuities in production hardware, whether detected by NDT, part failure, or destructive testing. This would begin to build an industry understanding of the types, sizes, and occurrences of physical discontinuities. While these would be fully pedigreed, the contributors would remain anonymous. In addition to a fee to support the maintenance, subscribers would also be required to contribute to the repository or certify a lack of physical discontinuities on a regular basis.
- Develop simple fatigue curves that compare the properties of smooth versus rough surfaces for common engineering materials that would give product teams a first-order approximation of the impact and need for performing NDT.

Industry Recommendations

The overall recommendations of this SAE EDGE™ Research Report can be summarized as follows:

1. The development and validation of methods to highlight potential discontinuities in radiographic examinations (e.g., DR, CR, CT). When used in combination with the employment of robotics to rapidly position and image parts, they will greatly decrease the time needed for an inspector to perform interpretation.
2. The development and validation of in situ and surface NDT methods for AM of polymeric composites. This will support the implementation of more complex geometries and structurally significant applications.
3. The development of affordable surface-finishing technologies that can address the wide range of geometries.
4. Research the impact of small, randomly distributed discontinuities on the fatigue and impact performance of a part. Since the majority of AM processes use solidification or sintering, small physical discontinuities will almost always be present. At some point, the size, distribution, and location of these transition from not impacting properties to having a significant detriment. A better understanding of this transition, combined with probabilistic methods, would enable organizations to develop ways of determining reliability for more critical AM parts.
5. Industry groups should initially focus on the material and process combinations that have the potential to make the largest impact. In metals, this would be laser powder bed fusion (L-PBF), followed by large puddle DED. In polymer composites, it would be ME, followed by L-PBF. In ceramics, this would be BJP.

Definitions

3D - Three-dimensional

AM - Additive Manufacturing

BJP - Binder Jet Printing

°C - Degrees Celsius

CAD - Computer-Aided Design

CR - Computed Radiography

CT - Computed Tomography

CAT - Computerized Axial Tomography

cm - Centimeter

DED - Directed Energy Deposition

DfAM - Design for Additive Manufacturing

DR - Digital Radiography

FR - Film Radiography

HIP - Hot Isostatic Pressing

kg - Kilograms

L-PBF - Laser Powder Bed Fusion

MHz - Megahertz

MJ - Material Jetting

mm - Millimeters

MPa - Megapascals

NDT - Nondestructive testing

PBF - Powder Bed Fusion

PoD - Probability of Detection

VP - Vat Photopolymerization

μm - Micrometer

Acknowledgments

Recognition should go first to all the participants, many of who also provided feedback on the draft version of this publication. Without their input and initiative, this SAE EDGE™ Research Report would not have been possible.

Anton du Plessis, Ph.D, *Stellenbosch University*
Michael Goerlik, Ph.D, *Federal Aviation Administration*
Dan Johns, *BEAMIT SPA Additive Manufacturing Technologies*
Mick Maher, *Maher & Associates LLC*
Ankit Saharan, Ph.D, *EOS*
Jim Williams, Ph.D, *The Ohio State University (Ret.)*
Dylan Yazbeck, *Jesse Garant Metrology Center*

The author of this document together with the SAE Team responsible for its creation join in expressing our deepest appreciation to all the individuals mentioned above.

Kevin T. Slattery D.Sc
The Barnes Global Advisors

Reference

1. Soja, M., De Silva, R., and Channa, R., "Surface Smoothing of Powder Bed Fusion Additive Manufactured Ti-6Al-4V Components," *Boeing Technical Journal*, Innovation Quarterly, Technical Articles, BTJ: A Smooth Finish, February 2019, https://www.boeing.com/features/innovation-quarterly/feb2019/btj-additive.page.

Contact Information

EDGEresearch@sae.org.

www.ingramcontent.com/pod-product-compliance
Lightning Source LLC
LaVergne TN
LVHW070937210926
848482LV00003B/267
9781468602401